Hiding in a Forest

Patricia Whitehouse

Heinemann Library
Chicago, Illinois

© 2003 Heinemann Library
a division of Reed Elsevier Inc.
Chicago, Illinois

Customer Service 888-454-2279
Visit our website at www.heinemannlibrary.com

All rights reserved. No part of this publication may be reproduced or transmitted in any form or by any means, electronic or mechanical, including photocopying, recording, taping, or any information storage and retrieval system, without permission in writing from the publisher.

Designed by Cherylyn Bredemann
Printed and bound in China by South China Printing Company
Photo research by Kathryn Creech

07 06 05
10 9 8 7 6 5 4 3

Library of Congress Cataloging-in-Publication Data

Whitehouse, Patricia, 1958-
   Hiding in a forest / Patricia Whitehouse.
     p. cm. -- (Animal camouflage)
   Summary: Describes how animals and insects living in the forest use
various forms of camouflage to survive, capture prey, and avoid predators.
   Includes bibliographical references (p.   ) and index.
   ISBN 1-40340-797-5 (HC), 1-40343-187-6 (Pbk)
   1. Forest animals--Juvenile literature. 2. Camouflage
(Biology)--Juvenile literature. [1. Forest animals. 2. Camouflage
(Biology) 3. Animal defenses.] I. Title. II. Series.
QL112 .W448 2003
591.47'2--dc21
                                                       2002010281

Acknowledgments
The author and publishers are grateful to the following for permission to reproduce copyright material: p. 4 Martial Colomb/PhotoDisc; p. 5 Geostock/PhotoDisc; pp. 6, 7 Oxford Scientific Films; pp. 8, 15, 18, 30 T Joe McDonald/Corbis; p. 9 Tom Brakefield/Corbis; p. 10 Alvin E. Staffan/Photo Researchers, Inc.; p. 11 Mark Smith/Photo Researchers, Inc.; p. 12 Jeremy Woodhouse/PhotoDisc; p. 13 Tim Zurowski/Corbis; p. 14 Richard A. Cooke/Corbis; p. 16 Tom J. Ulrich/Visuals Unlimited; p. 17 Jim Zipp/Photo Researchers, Inc.; p. 19 Hal Horwitz/Corbis; p. 20 Bill Beatty/Visuals Unlimited; p. 21 Ralph A. Clevenger/Corbis; pp. 22, 23 Michael & Patricia Fogden/Corbis; p. 24 David M. Schleser/Nature's Images, Inc./Photo Researchers, Inc.; pp. 25, 30B John Mitchell/Oxford Scientific Films; p. 26 Tom & Pat Leeson/Photo Researchers, Inc.; p. 27 Manfred Danegger/Photo Researchers, Inc.; p. 28 J. Serrao/Visuals Unlimited; p. 29 Jeff Lepore/Visuals Unlimited.

Cover photography by Tom & Pat Leeson/Photo Researchers, Inc.

Every effort has been made to contact copyright holders of any material reproduced in this book. Any omissions will be rectified in subsequent printings if notice is given to the publisher.

Some words are shown in bold, **like this**. You can find out what they mean by looking in the glossary.

To learn about the ermine on the cover, turn to page 26.

Contents

Hiding in a Forest 4
Hiding in a Tree 6
Hiding on the Ground 8
Hiding in a Forest Lake 10
Hiding by Looking Different 12
Hiding to Hunt 14
Different Ways to Hide 16
Pretending to Be Part of the Forest 18
Pretending to Be Another Animal 20
Tricking with Color 22
Hiding in Plain Sight 24
Changing Colors 26
Surprise! 28
Who Is Hiding Here? 30
Glossary 31
More Books to Read 32
Index 32

Hiding in a Forest

Many animals live in the forest. Some forest animals are hard to see. They use **camouflage** to help them hide.

Some animals hide so they do not get eaten. Others hide from animals they want to catch and eat. There are many ways to hide. This katydid has **cryptic coloration**.

Hiding in a Tree

This peppered moth is the same color as the tree. This makes the moth hard to see. Animals that look like their **habitat** have **cryptic coloration**.

The peppered moth is easy to see here. It is on a dark-colored tree. Birds and other animals can find it and eat it.

Hiding on the Ground

Animals also hide from **predators** on the ground. The ruffed grouse lays its eggs in the leaves. The grouse and its nest have **cryptic coloration**.

In winter, you can see the grouse right away. The grouse's **camouflage** colors do not help it hide in the white snow.

Hiding in a Forest Lake

This minnow has **cryptic coloration**. It lives in a forest lake with a rocky bottom. It is the same color as the rocks on the bottom of the lake.

When they are not on the rocky bottom, minnows are easy to see. Big fish and other **predators** can find them and eat them.

Hiding by Looking Different

Some animals have color **patterns** that break up their shape. This is called **disruptive coloration**. This **fawn's** spots look like sunlight coming through the trees.

This nighthawk has stripes on its feathers. The stripes make it hard to see the whole nighthawk.

Hiding to Hunt

Some animals use **disruptive coloration** to help them hunt. The body of this timber rattlesnake has dark and light colors. This makes the snake hard to see.

The spots on the bobcat's fur help it hide. The spots let the bobcat get close to rabbits, squirrels, and other animals. Then it can catch them and eat them.

Different Ways to Hide

Sometimes saw-whet owls hide in holes in trees. This makes the saw-whet hard to see.

Saw-whet owls can also hide on tree branches. They are still hard to see because they have **cryptic coloration.**

Pretending to Be Part of the Forest

This praying mantis is hunting for food. It is using a kind of **camouflage** called **aggressive mimicry**. To trick its **prey,** the mantis makes itself look like a plant.

This praying mantis has caught a butterfly. The butterfly did not see any danger until it was too late.

Pretending to Be Another Animal

A bee can sting. Its yellow and black stripes tell **predators** to stay away.

This bee fly cannot sting. But it **mimics** a real bee. It is yellow and black like a bee. Predators leave the bee fly alone.

Tricking with Color

Coral snakes are **venomous.** They also have red, white, and black stripes. **Predators** stay away from coral snakes.

King snakes are not venomous. But they have red, white, and black stripes. King snakes **mimic** coral snakes. Predators will also leave king snakes alone.

Hiding in Plain Sight

Some forest animals do not hide. They **mimic** things that **predators** do not eat. This swallowtail caterpillar looks like a bird dropping. No animal wants to eat bird poop!

This is not really a stick. It is an **insect** called a walking stick. It uses mimicry to look like a stick swinging in the wind.

Changing Colors

Some forest animals change colors. In the summer, this ermine's brown fur looks like leaves on the forest floor. This helps the ermine hide from **predators**.

26

In the winter, there is snow on the ground. The ermine's brown fur would not help it hide. So the ermine grows white fur to hide in the snow.

Surprise!

Some forest animals scare away **predators**. This polyphemus moth is resting on a tree. Right now, its wings are closed.

Here the moth is showing the **eyespots** on its wings. These spots look like owl eyes. A scared predator will leave this moth alone!

Who Is Hiding Here?

What animals are hiding here?
What kind of **camouflage** do they have?

For the answer, turn to page 8.

For the answer, turn to page 25.

Glossary

aggressive mimicry   animal looks and acts like a plant or animal that is not dangerous

camouflage   use of color, shape, or pattern to hide

cryptic coloration   colors that make an animal look like the place where it lives

disruptive coloration   pattern of colors on an animal that makes it hard to see the whole animal

eyespots   spots shaped like animal eyes that scare predators away

fawn   young deer

habitat   place where an animal or plant lives

insect   animal with wings, six legs, three body parts, and a hard shell

mimic, mimicry   one animal looks and acts like a plant or another kind of animal

pattern   colors arranged in shapes

predator   animal that eats other animals

prey   animals that are eaten by other animals

venomous   animal that puts a dangerous liquid into another animal

More Books to Read

Arnosky, Jim. *I See Animals Hiding.* New York: Scholastic, Incorporated, 2000.

Galko, Francine. *Forest Animals.* Chicago: Heinemann Library, 2002.

Kalman, Bobbie. *What Are Camouflage and Mimicry?* New York: Crabtree Publishing Company, 2001.

Index

aggressive mimicry 18–19
bee 20
bee fly 21
bobcat 15
camouflage 4–5, 9, 18,
caterpillar, swallowtail 24
cryptic coloration 6–7, 8–9, 10–11, 17
disruptive coloration 12–13, 14
ermine 26, 27
eyespots 29
fawn 12
katydid 5
mimic, mimicry 18–19, 20–21, 22–23, 24–25

minnow 10, 11
moth
   peppered 6, 7
   polyphemus 28, 29
nighthawk 13
owl, saw-whet 16, 17
praying mantis 18, 19
predator 8, 11, 20, 21, 22, 23, 24, 26, 28, 29
prey 18
ruffed grouse 8, 9
snake
   coral 22, 23
   king 23
   timber rattle 14
walking stick 25